Foreword

This book is intended to present and promote a new method of propulsion that will permit spacecraft in micro-gravity to accelerate for extended periods of time thus allowing the craft to obtain a very high velocity without expelling mass. Also since the propulsion system itself uses MHD technology and requires no moving parts the craft can be expected to have extremely long life. Long life is necessary for long term deep space probes (in excess of 35 years), such as possible probes to another solar system.

I will illustrate that it is possible to accelerate a closed system, such as a vessel containing a fluid; the fluid's momentum is dismembered and redirected by canceling certain unwanted forces through a geometric structure. The forces within the geometric structure are constructed so that both static and dynamic forces are created. The rigid static forces result in the dismemberment of the fluid's momentum while the dynamic forces add to produce composite propulsion. The thrust or propulsion is limited only by the number of vessels their power source and the engineered efficiency of the overall system.

The books material is divided into two parts, the main part or core material and the appendices. The core material is most suited for the reader who is interested in the "big picture". The appendices are composed of proofs, thought experiments, etc. for students of engineering, physics or those interested in more details.

If desired the main part should be read first and the appendances or proofs can be read later for these who are interested in the detailed physics and mathematical proofs. The appendices require knowledge of high school physics and algebra, although no calculus is necessary.

It should be understood that in this book Newton's third law is adopted to a the 21'st century interpretation format "An object that is in motion will not change its velocity unless an unbalanced force acts upon it or within it.", instead of, "An object that is in motion will not change its velocity unless an unbalanced force acts upon it". The first interpretation is used in order to advance beyond the 17th century.

It is believed by many engineers that an open system (some form of rocket) is the only means of space travel in micro-gravity. It should be apparent by anyone who has common sense that if Newton had known it was possible to have a flying machine inside a closed system such as a toy helicopter inside a Styrofoam box he would have included the working of a closed system in his theory. It is fairly certain to say that Newton was not thinking about having objects inside of objects, he was thinking instead about forces acting on a singular object. And it is almost certain that he was not thinking about having fluids acting inside of objects with technologies that didn't exist at that time. Sometimes things "fall through the cracks" and that is what has almost certainly happened with this invention.

The material presented is dedicated to a form of propulsion in a closed system by changing the direction of the mass and or the dispersion of fluid and the cancellation of forces within a geometric structure. It is dedicated to near light or constant acceleration to the point of near light speed. No attempt is made to suggest that faster than light speed is not possible with some sort of future technology, but the technology of this material is limited to what is known at this time and does not take into consideration gravity waves, dark energy, or the bending of space.

This material is designed to present a faster approach to space travel in micro-gravity. At the same time the limitations of an open system (any type of rocket) in micro-gravity are discussed in detail.

Whereby the limitations of an open system are discussed in micro-gravity, It should be understood that there is a definite need for open systems because of their extreme power in a low velocity environment, which makes them necessary to escape the gravity of a planet. They are to be used in conjunction with the closed loop system described in this material.

Closed Loop Space Propulsion New Faster Approach

Constant Acceleration Propulsion Using Inertia

Constant acceleration propulsion could be the single most important thing in outer space travel and theoretical research in the future. In the past, constant acceleration was not necessary or needed because the tremendous velocities and atmosphere of the planet would disallow this phenomenon. However in outer space, constant acceleration to extremely high sub light velocity in a relatively short period of time is possible.

Much work will be done to research and develop constant accelerant technology. In this book I will show and explain the operation of two constant acceleration constructs. One of these constructs is a very crude mechanical device and is used for purely academic reasons; the other is a practical closed propulsion system and will be explained in detail.

It is important to understand the force of a system and why a special force will allow us to travel at incredible velocities in outer space. This is a propulsion system that uses principles and or forces that are seldom observed, or may be taken for granted in the everyday world. This makes it difficult to conceive of their mode of operation. However, I believe I will provide enough information to convince a reasonable person with moderate background that there exists a unique working propulsion system that offers many advantages over other forms of space propulsion. The only other form of propulsion through space is by the means of rockets. These are in the form of three well known types; chemical, plasma and ion.

The special force I am proposing for this new type of propulsion is sometimes referred to as G-force and is actually a type of directional and inertial force or combination of inertial directional vectors that result in a directional dynamic force. The G-force or inertial force may have greater value in the field of space propulsion than previous believed. One inertial force that goes unnoticed is centrifugal force. A centrifugal force uses a force that is very similar to gravity, for instance; the vector sum of the forces of a gyroscope result in a non-directional static force that resists a change in the direction of motion. According to Albert Einstein, acceleration type forces G-forces or centrifugal forces are similar to or even equivalent to gravity.

For instance, "Einstein's principle of equivalence', says that if you're on an elevator accelerating there would be no way you could determine the mass of the planet you were on. He said, if you were to move a laboratory on the elevator with you, there would be no experiment you could perform on the elevator to tell a difference between gravity caused by the elevators acceleration or the mass of the earth.

Using Einstein's basic equation in his special theory of relativity it is a fact that gravity will accelerate a mass up to approximately .8 times the speed of light with a constant acceleration. At this point mass begins to dilate to infinity thus slowing the acceleration of the mass of the body. As the speed of the mass approaches light, acceleration of the mass approaches zero. The construct in this book for space travel, will not begin to slow in acceleration until reaching approximately 90 percent the speed of light, but will never reach the seed of light. Light travels approximately 600,000,000 miles an hour. Today's latest ion and plasma rockets approach a maximum speed of 60,000 miles per hour. Even though they may use powerful super conducting magnets, the exhaust velocity limits their velocity through the law of conservation of momentum, a short coming of open propulsion systems. This is not true with a closed system. The latter is in accordance with Einstein's principle of equivalence, and since, no mass flows out of the system, the law of conservation of momentum is not being used. However, high power at relatively low velocity is needed to break free of the earth's gravity and open systems are required to reach orbit.

It is possible that by using super conducting magnets, inertia driven systems such as those described in this book could produce the same if not greater force than chemical rockets; however, an impractical amount of electrical power would be required. Chemical rockets have their maximum force at the beginning of takeoff and almost no force at approximately 20,000 miles per hour. At this speed more fuel would have little or no effective thrust.

In accordance with the law of conservation of momentum ($m1*v1 = m2*v2$). Where m1=mass of the exhaust gas, m2=mass of the rocket, v1=velocity of the exhaust gas and v2= the velocity of the rocket the mass. Of course the mass of the exhaust gas m1, m2, and v1 are fixed, therefore, when v2 reaches a fixed force to balance the equation there will be no further increase in the velocity of the rocket. Nick Herbert PHD in his book <u>FLT Loopholes In</u> <u>Physics</u> said, "An effective limit on a rocket's speed is set by its own exhaust velocity".

Creating Inertial Force

The G-force cause by the acceleration of an elevator is a form of centrifugal force but it is usually called inertia when the force acts in a straight line. However, they both produce the same results that of a gravitational force or of inertia/G-force exerted on a mass changing direction.

There are two conceivable ways to create this artificial gravitational force. One way is to continuously increase the velocity of a mass. You have probably noticed this when driving your car. An increase in velocity will cause a gravitational type force as you are pushed back against the seat, sometimes referred to as a G-force.

The other way to create the same type of force is simply to continuously change the direction of a mass. The force produced by changing the direction of the mass is proportional to the rate of change of direction and the mass changing direction.

If you use the first method of continuously increasing the velocity of the mass there is a limit in which you can increase velocity to achieve these types of gravitational forces. This is similar to the type of system that is used in all rockets; chemical, plasma and ion no matter how fast the gasses, plasma or ion mass can be pushed out of the system.

The maximum amount of force or thrust of a rocket occurs during the first few seconds of its operation. A large rocket such as the Space Shuttle will expel a large amount of mass during the first phase of its flight, resulting in a temporary increase in acceleration. This is true because the rocket's force is acting on a lighter structure that is less impeded by the thinning air of the upper atmosphere. From that point on the thrust will begin to decrease to zero. The thrust will become zero when the exhaust velocity of the rocket is equal to the rocket speed itself and causes no further increase in the velocity of the rocket.

This seems to satisfy the present need for orbital propulsion and is probably because the present system works well enough (satellite operation for 15 years). Advances have been made but only of an evolutionary nature.

In Fig. 1 it can be seen that with even a high initial acceleration it's only a matter of time before the constant acceleration of Fig. 2 exceeds the velocity of the rocket in Fig. 1.

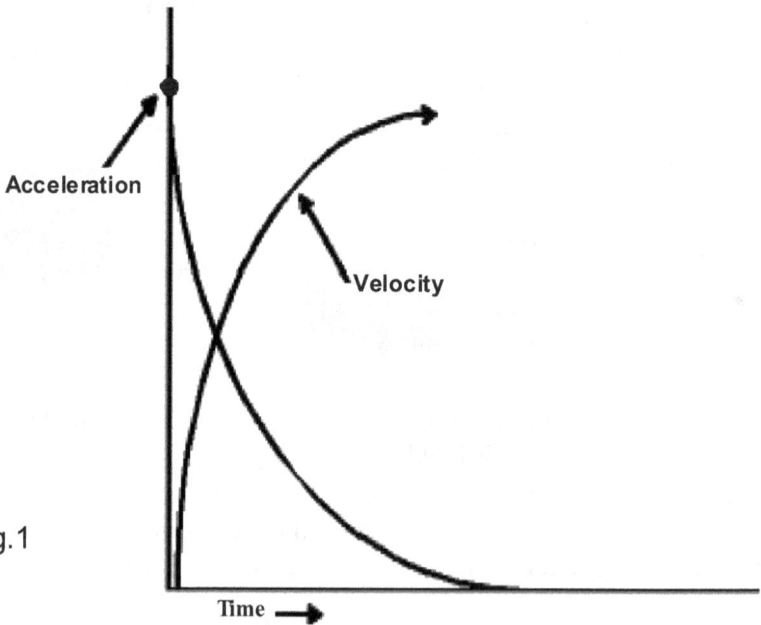

Acceleration

Velocity

Fig.1

Time

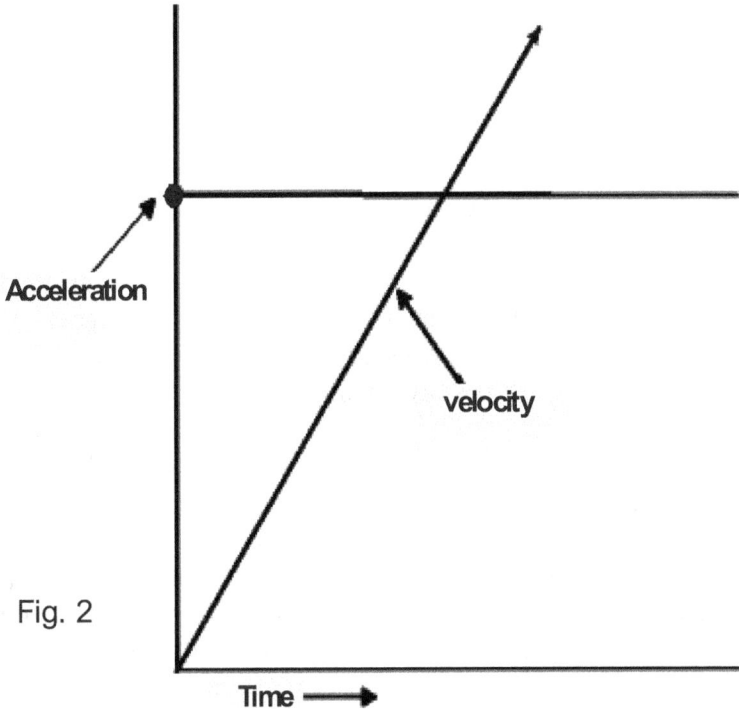

Fig. 2

The only way to sustain a constant acceleration is to change the direction of a moving mass. The mass can be solid, liquid, or gas. However with a gas, the velocity of the moving gas would have to be extremely high to have any effect.

Mechnical unidirectional gyro force construct

A mechanical or constant acceleration engine, "gyro engine" is shown in. Fig. 3. The unidirectional gyro force is produced from the Sliding Rings. As each rod spins, the velocity of the Sliding Ring mass is changed by changing the distance of the mass along the length of the rods. A motor is attached to the rods to supply the necessary torque. The "pancake" electromagnets are supplied energy through the comutaters of the rotory shaft. The timing of the electromagnets is controlled by the necessary electronics.

To balance the motion of the rotating rods, a symetrical unit is placed on the bottom side of the top structure. The rotor on the opposite side will spin in the opposite direction. This is necessary to prevent the motor from chasing the rotor. The operation is such that the mass of the top will change position in sycronization with the bottom mass, but in the opposite direction.

This type of machine has limits as to the amount of force it can produce. The limis are determined by the spin rate and the rate at which mass can be moved along the spinning rods and the mass of the Sliding Rings. However, it is unlikely that 1g of acceleration could be obtained. It would be necessary for the unit to create 1 G-force of acceleration to lift it's own weight.

Nick Herbert PHD in his book FLT Loopholes In Physics made the point that 1 g of acceleration, taking into account Einstins's theory of relativity (relativity has practically no effect at velocities less than 95% of light speed), will allow a space ship to reach near the speed of light in one year. However, no inerital device can have constant acceleration near the speed of light, due to the Einstein limit. An acceleration of 1 g would take only 2 years to reach over 92% the speed of light. Chemical ion, plasma (open systems) rockets using the largest possible initial force (or thrust), with an infinate amount of time, would not result in near light speed. This is because they are open systems and use the conservation of momentum for their operation.

The device in Fig.3 is very crude, and used for academic purposes only. In order for it to produce 1 g of acceleration it would have to lift it's own weight, for practical purposes this is highly unlikely. Producing 1 g requires that the device lift it's own weight, the power source, shell and any insurments required for a probe. Of course if it were to be used for maned space travel it would require much more space than the bare bones system and chemical fuel in order escape the earths gravity. The system would also suffer from wear and tear and is not suited for the long term of space travel. There is also the problem of lubrication due in part to the extreme temperatures of space.

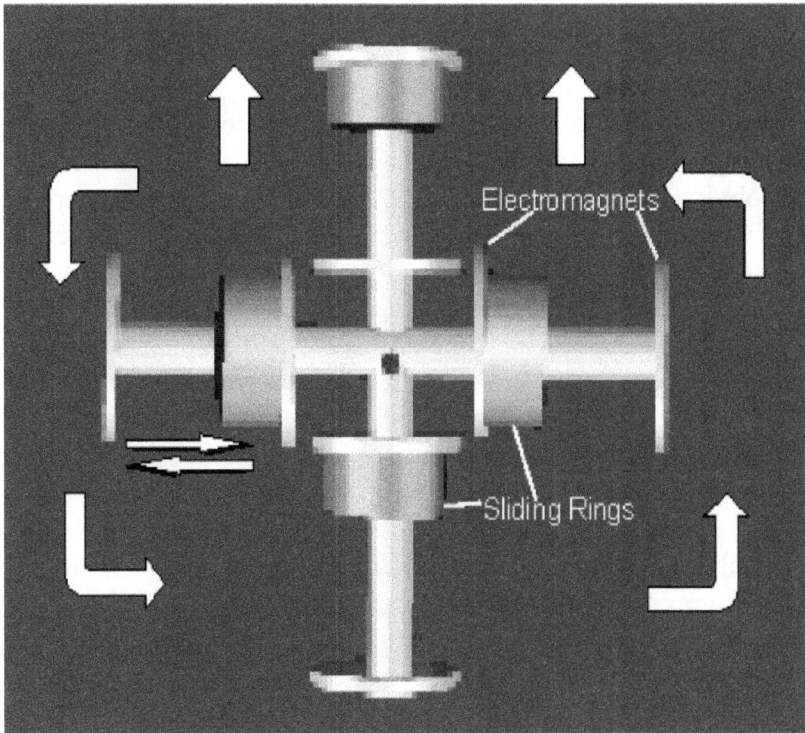

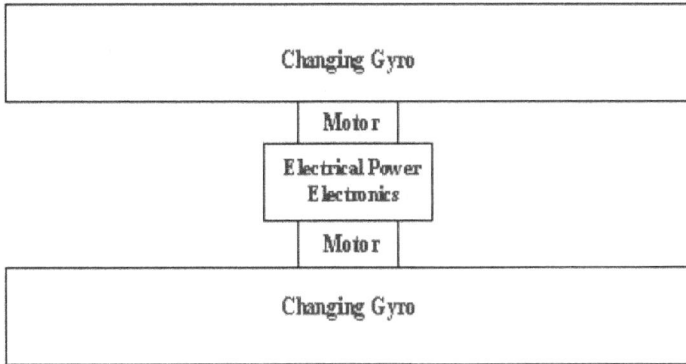

Fig. 3

The system in Fig. 3 deinstrates the use of two spinning gyros that change shape in order to demonstrate a dynamic force within a closed system.

How Inertial Devices Can Travel Near Light Speed
With Constant Acceleration

The obvious answer is that since everything moves along with the system and the velocity of the system does not affect the propelling force, it is obvious there is constant acceleration. The other answer involves the concept of independence, in which the action of only one force acting against a single mass can create a movement of the mass itself. All forces in nature except gravity involve the interaction of at least two forces for movement. This includes a rocket that pushes against its expelling matter or a car tire pushing against a road. In nature, any action involving two forces and two masses cannot have constant acceleration (or increased acceleration) over time; also modern systems for space require some type of mass fuel to be expended

The only thing that can have constant acceleration is a single force, such as gravity acting on a mass. Gravity is the only example found in nature. For example, without an atmosphere, objects fall to the earth at a constant acceleration of 32 feet per Second Square. Of course this means that every 32 feet the objects will double its velocity. You can see the tremendous velocity that can be obtained from constant acceleration in a relativity short period of time.

Using gravity as the standard for constant acceleration and combining that with Einstein's principle of equivalence concludes that any inertial type force or unidirectional gyroscopic force will have a constant acceleration in outer space. The constructs I describe use a directional gyroscopic force or inertial force or some other type of independent force which when acting within a structure incur a gravitational propulsion or independent dynamic force.

Practical Liquid Metal Construct

A practical construct is obtained by accelerating a liquid metal through a geometric structure. An MHD accelerator/pump is used to transform electrical energy into kinetic energy.

The MHD accelerator uses the Lorenze Force. The Lorenze Force is not a magnetic force, that is, it is not a force that results from the interaction of two magnetic fields, but rather a magnetic force that acts on electrons or charges within the conducting liquid. The Lorenze Force will act on any liquid that is conductive, however the best choice is a liquid that is both heavy and has a very high conductivity, such as a heavy liquid metal. Practicable propulsion is achieved by using heavy liquid metals.

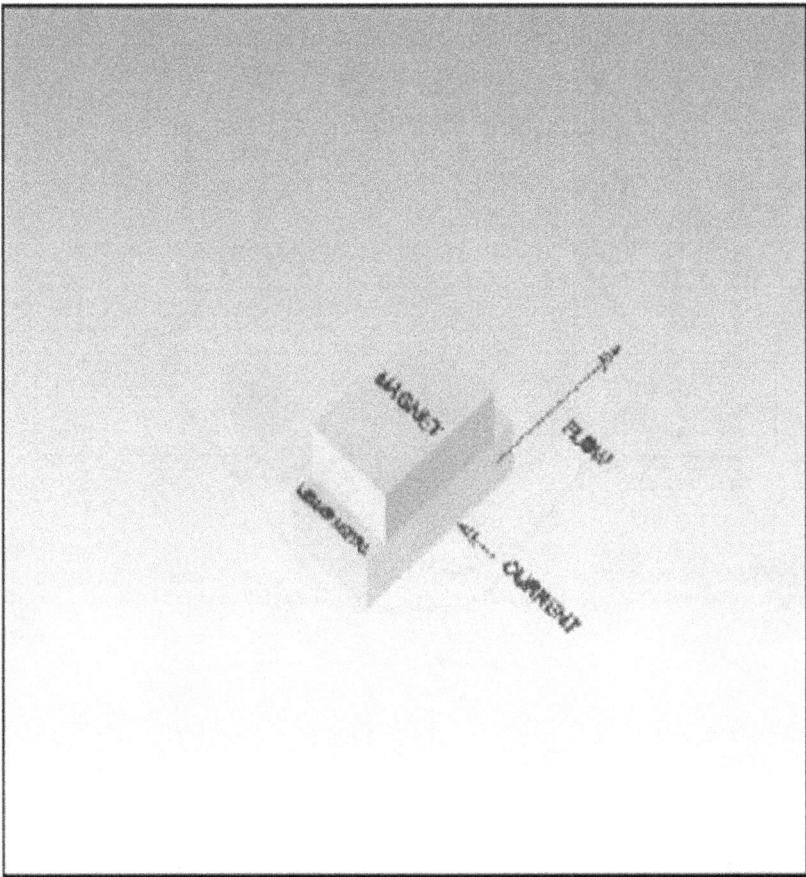

Figure 4

Figure 4 demonstrates the three dimensional interaction of the current, magnetic field, and the flow of the liquid metal. Using the MHD accelerator in fig 4 and combining it with a geometric structure gives the Closed System Propulsion Using Liquid Metals shown in fig. 5. See **Appendix C** for more details on the MHD accelerator.

As can be seen in Figure 5 due to the flow of the liquid metal all horizontal forces cancel while the vertical forces will add collectively, thus allowing propulsive force.

Representation of Forward Propulsion
Using Concept From Appendix A and B

Reversing The Direction of Current Flow Will
Allow a Forward or Backward Dynamic Force

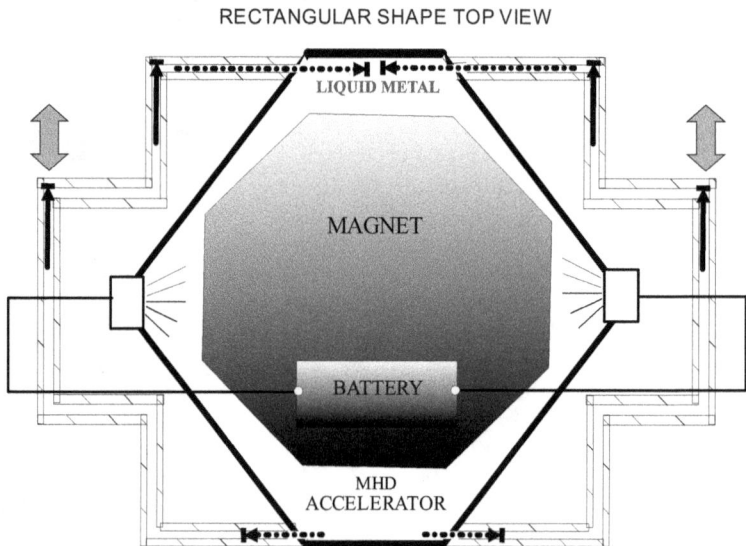

RECTANGULAR SHAPE TOP VIEW

LIQUID METAL

MAGNET

BATTERY

MHD
ACCELERATOR

Figure 5

This Basic Closed System Propulsion Construct is the building block for all space propulsion systems in this book. **See Appendix A and B** for an in depth explanation of the Basic Closed System Propulsion Construct. A Closed Propulsion Engine can be made one or more of these Closed System Propulsion Constructs. The system in general will be referred to as a Closed Loop propulsion unit or Closed Loop Propulsion component.

Advanced Closed Loop Propulsion System

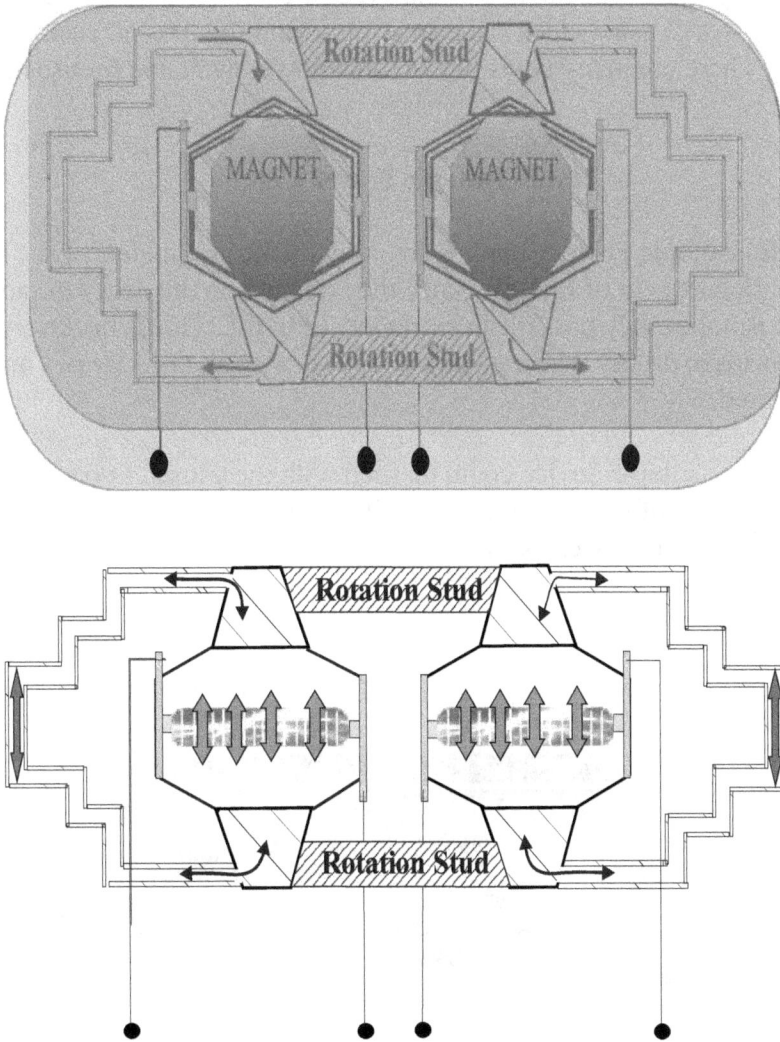

Figure 6

Figure 6 shows the advanced Closed Loop propulsion engine. For future reference it will be referred to as the Closed Loop Propulsion Engine. For operational details of the Closed Loop Propulsion Engine refer to **Appendix C**.

Power Sources for the Closed Loop Propulsion Engine

One tremendous advantage of this type of closed loop propulsion is that it works using electrical energy only, and requires no chemical fuel. There are many types of electrical power available for use in space.

Radio Isotopic Power Cell

Radio Isotopic Power Supplies are allowed for use in outer space by The Department of Energy. Since they operate on the decay of radioactive isotopes they possess an extremely long life. This is necessary in deep space probes where electrical power in excess of 35 years may be required.

A Radio Isotopic device could be used with the Closed Loop Propulsion Engine of Figure 6 for a versatile number of applications.
A Radio Isotopic Cell is shown below in the Figure below.

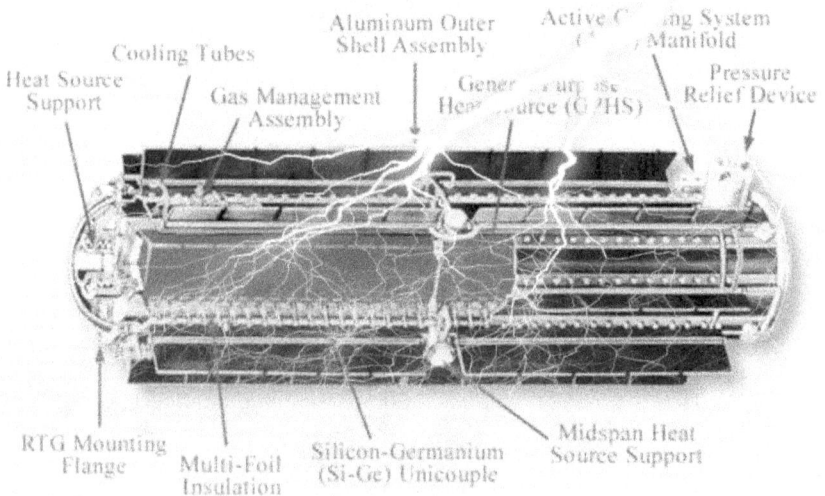

(European Space Agency/European Space Research and Technology Center) explains the RTG technology as follows:

"What Are RTGs?

RTGs are lightweight, compact spacecraft power systems that are highly reliable. RTGs are not nuclear reactors and have no moving parts. They use neither fission nor fusion processes to produce energy. Instead, they provide power through the natural radioactive decay of plutonium (mostly Pu-238, a non-weapons grade isotope). The heat generated by this natural process is changed into electricity by solid-state thermoelectric converters. RTGs enable spacecraft to operate at significant distances from the Sun or in other areas where solar power systems would not be feasible. In this context, they remain unmatched for power output, reliability and durability.

Safety Design
More than 30 years have been invested in the engineering, safety analysis and testing of RTGs. Safety features are incorporated into the RTG's design, and extensive testing has demonstrated that they can withstand physical conditions more severe than those expected from most accidents.

First, the fuel is in the heat-resistant, ceramic form of plutonium dioxide, which reduces its chance of vaporizing in fire or reentry environments. This ceramic-form fuel is also highly insoluble, has a low chemical reactivity, and primarily fractures into large, non-respirable particles and chunks. These characteristics help to mitigate the potential health effects from accidents involving the release of this fuel.

Second, the fuel is divided among 18 small, independent modular units, each with its own heat shield and impact shell. This design reduces the chances of fuel release in an accident because all modules would not be equally impacted in an accident.

Third, multiple layers of protective materials, including iridium capsules and high-strength graphite blocks, are used to protect the fuel and prevent its accidental release. Iridium is a metal that has a very high melting point and is strong, corrosion resistant and chemically compatible with plutonium dioxide. These characteristics make iridium useful for protecting and containing each fuel pellet. Graphite is used because it is lightweight and highly heat-resistant."

Using Fission Reactors in Space

Fission reactors have been discussed for use in outer space, but due to their potential for failure and risk of neutron radiation they are generally considered untenable. These would be closed system reactors with extremely high power and are not required for this Closed Propulsion System.

Solar Arrays

Because of other planets distance from the Sun, solar arrays are not feasible power sources for spacecraft. Solar arrays for orbital craft however are well suited for the Closed Loop Propulsion Engine guidance correction thrusters.

Applications for the Closed Loop Propulsion Engine

Satellites and Station Placement

Vehicles operating in outer space must typically carry all their fuel in order to accelerate, decelerate, and steer the vehicle. This fuel may include a solid or liquid propellant, including, for instance, liquid oxygen and liquid hydrogen, and this fuel is consumed until eventually there is no more fuel to burn and the vehicle is left adrift. If the vehicle is in orbit around the earth, its orbit may decay until eventually the vehicle falls back toward the planet, typically burning up in the atmosphere along the way.

Current thrust methods for satellites and "station placement" (also referred to as attitude adjustment thrusters, used not only to orient the vehicle but also to "boost" its orbit) may use liquid propellant as fuels or may even be as unsophisticated as simply blowing nitrogen gas from a tank through a nozzle. Relatively new technology has introduced ion engines which accelerate atomic particles to extremely high speeds to create thrust more efficiently. An ion engine may use heavy but inert Xenon gas as fuel, as well as electricity from large solar arrays. The thrust comes from accelerating the ions (positively charged atoms) and expelling them at very high speeds.

In all of the above cases, once the fuel source carried with the vehicle is depleted, there is no longer any ability to control the movement of the vehicle. There is a definite need for a propulsion system which draws its energy from a source, such as electricity, which can be provided by means of solar panels.

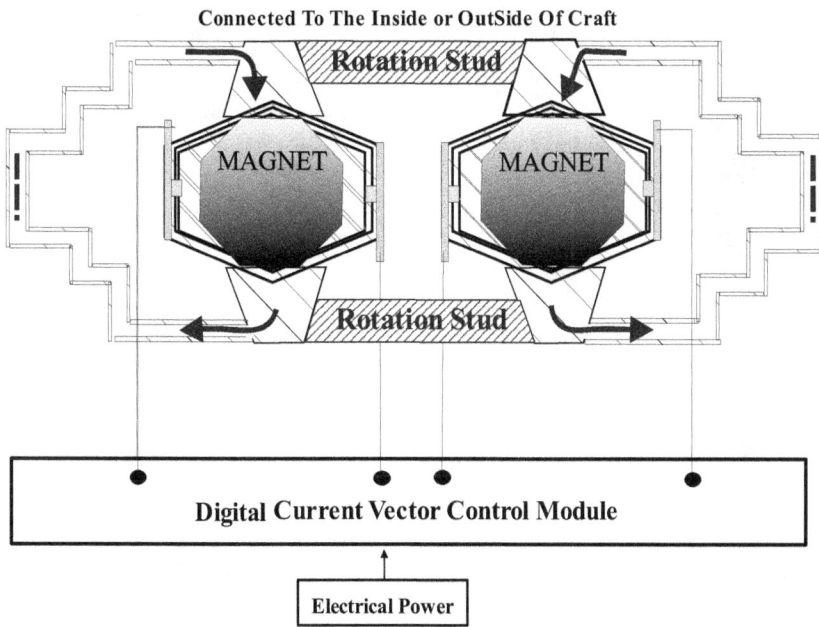

Figure 7

Figure 7 describes a computer position controlled Closed Loop Propulsion System. Electrical power is supplied to a digital current module. The digital current module supplies correction current to the multi-directional Closed Loop Propulsion Engine which in turn corrects the position of the craft.

The rate of correction would depend upon the engineering of the closed propulsion system the number of closed propulsion system modules used and the power supplied by the correction circuits to the close propulsion system Figure 8 illustrates but one type of possible satellite construct using the closed propulsion system.

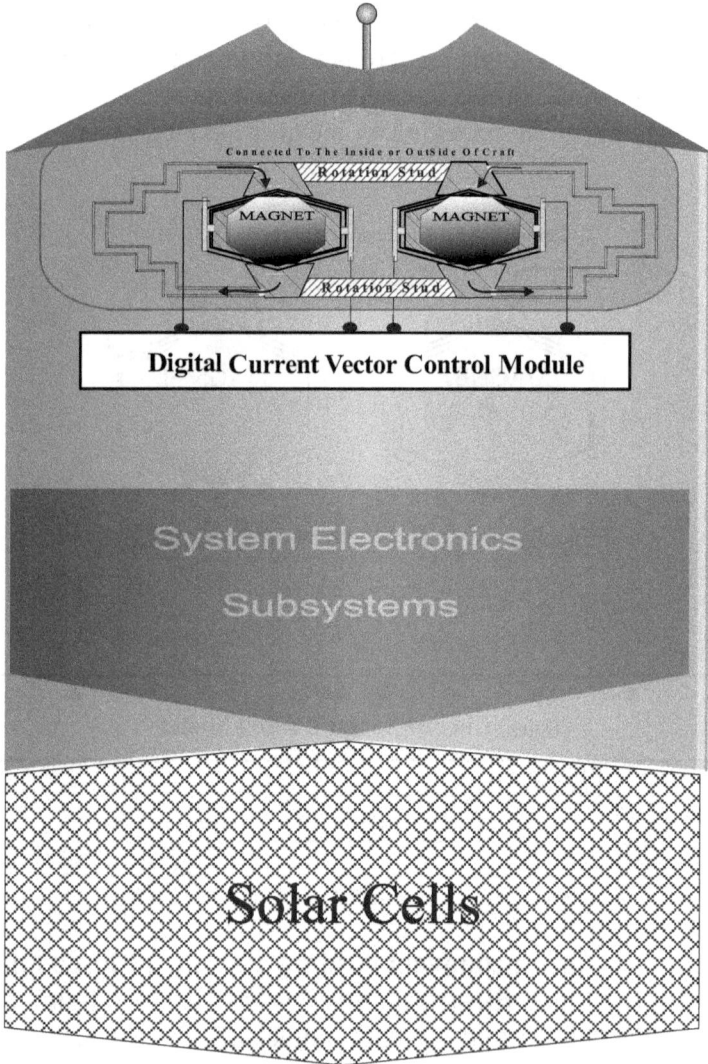

Figure 8

Manned Travel to Mars

The following is from **YouTube**
http://www.youtube.com/watch?v=Zj53rVWK5z0&feature=r
elated **www.adastrarocket.com**

"An electric propulsion thruster such as VASIMR would provide a faster travel time compared to using chemical rockets. The VASIMR engines would speed up the spacecraft towards Mars and then slow down the spacecraft once past the half way point from Earth to Mars. Ion engines in general provide a low amount of thrust (up to several Newtons or tens of Newtons), but do so for many hours or months. In this way, a spacecraft with a given mass can continually speed up and attain velocities much higher than with chemical rockets. A mission to Mars using plasma propulsion would also have an abort capability should something go wrong early on in the mission, chemical rockets do not provide for this contingency."

One important application for the Closed Loop propulsion engine is manned travel to Mars. The mission would in general be the same as described above, except that since it is a closed system it would not attempt to "strong arm" conservation of momentum. As described earlier Closed Loop propulsion or closed loop fluid drive uses a system of canceling fluids and aiding fluids to incur a series of independent dynamic forces which result in a singular dynamic propulsive force.

Using the law of conservation of momentum with an extreme method requires that mass is expelled at tremendous velocity. Expelling matter, even extremely low density matter such as plasma at such velocity, requires enormous amounts of electrical power not to mention the amount of liquid hydrogen required. Upon reflection it is apparent that closed loop propulsion is the only viable means for this mission. Using Closed Loop propulsion the fuel necessary for the required assent from Mars (liquid hydrogen and oxygen) could also be used to generate electrical power for the Closed Loop propulsion

Interstellar Travel

One important application for the Closed Loop propulsion engine is interstellar space travel. Because of the extremely long time required even for the closed loop system described in this book a space probe is necessary. A space probe is a scientific space exploration mission in which a robotic spacecraft leaves the gravity of Earth and travels to interstellar space.

Constant acceleration over extend periods of time will result in near light speed, but even near light speed is extremely limited compared to the dimensions of outer space. Therefore correct choice of which one of the closest solar systems is critical for the success of a mission.

The Astronomical Society of the Pacific reports
"The closest stars to our Sun are in the three-star system called Alpha Centauri, a popular destination for interstellar travel in works of science fiction. UCSC graduate student Javiera Guedes used computer simulations of planet formation to show that terrestrial planets are likely to have formed around the star Alpha Centauri B and to be orbiting in the "habitable zone" where liquid water can exist on the planet's surface. The researchers then showed that such planets could be observed. To study planet formation around Alpha Centauri B, the team ran repeated computer simulations, evolving the system for the equivalent of 200 million years each time. Because of variations in the initial conditions, each simulation led to the formation of a different planetary system. In every case, however, a system of multiple planets evolved with at least one planet about the size of Earth. In many cases, the simulated planets had orbits lying within the habitable zone of the star."

A closed system space probe craft could reach Alpha Centauri B and start sending back pictures to Earth in as little as 30 years. The craft could use artificial intelligence to search for planets around the star, it could then send back detailed pictures which could reveal possible life on other planets. These of course would not be merely microscopic evidence but could show things such as water, trees, animals, houses, and possibly advanced structures. All this could be known as early as 30 to 40 years depending upon the engineering of the probe and of course its power source and other particulates. The Figure below shows a crude design of such a probe.

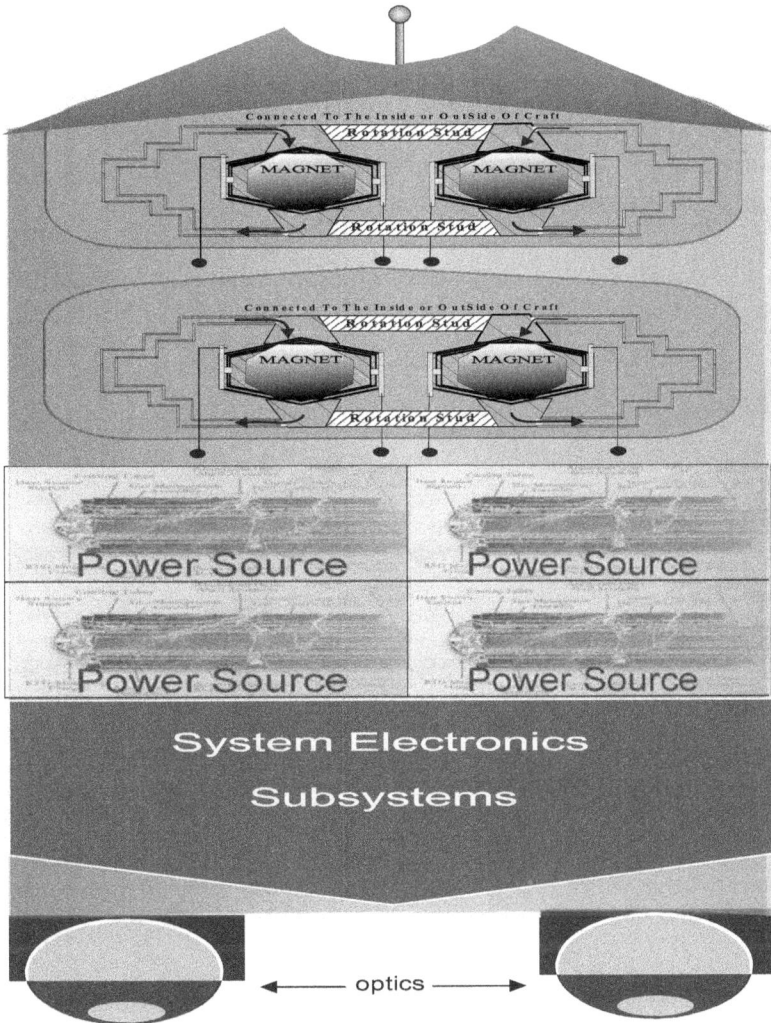

Figure 9

Proof of Closed System Propulsion Construct
Using Liquid Metals Part 1

Change in mass direction can be used to induce dynamic forces in a geometric structure. The basic components are, a shell geometric structure, power source and a fluid mass, preferably a heavy liquid metal or metal alloy.

Figure A-1 is an example of a liquid metal that is being circulated at a velocity Vx, by means of steady state pump. Power is supplied to the pump by an external power source, not shown.

The liquid metal is circulated through tubing that is firmly secured to a scale. For practical purposes the scale is firmly secured to a stationary structure so the weight of the tubing and pump can be measured.

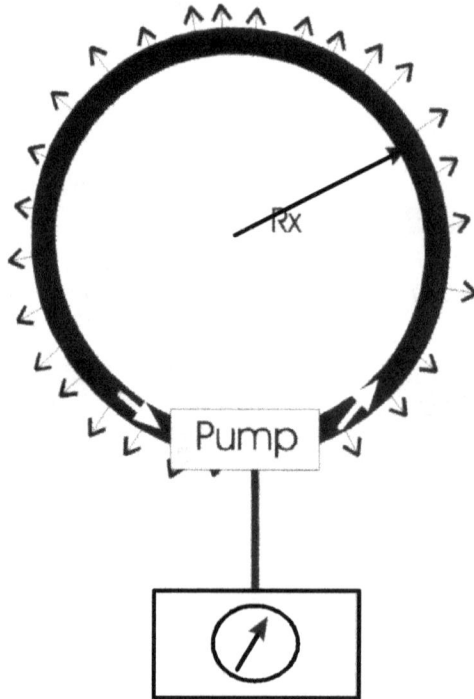

There is a centrifugal force from the circular motion of the liquid metal directed toward the center of the tubing proportional to the velocity of the liquid metal. Since the liquid metal is symmetrically flowing about the radius of the tubing, All units of arc length are equally acted on by a centrifugal force pulling at the center of the radius proportional to equation (1). M is unit mass per time, Vx is velocity of the liquid metal, and Rx is the radius of the tubing construct.

(1) $F=M(Vx^2)/Rx$

The force in figure A-1 will always measure the same weight no matter what the velocity of the liquid metal Vx. This is because all of the forces are equal and opposite around the circular construct.

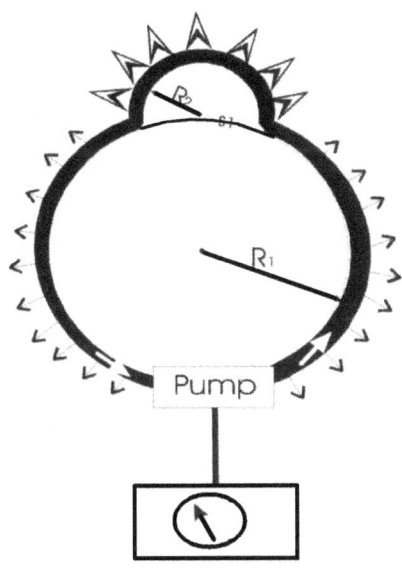

A-2

In figure A-2, a section of tubing is removed equal to s1, A small circular construction approximately the diameter of s1 is inserted. This circular construction is being used to demonstrate the effect of geometry in producing desired selective force from a flowing fluid.

As you can see in Figure A-2, there are much larger force represented by the half circular-tubing construct, as opposed to the force acting on the arc length s1. This is due to the smaller radius of r2 which results in a much greater vector force, resulting from the vector forces of the liquid metal.

(2) Force of half circular-tubing construct = M(Vxsquared)/r2

Using equation (1), it can be seen that the force per unit length/per second is M(Vxsquared)/Rx and the force from the smaller half-circular construct is per unit length/pre second is M(Vxsquared)/r2. It should be noted that the arc length of the tubing construct is much larger than the removed section sl. Also, the radius of the tubing construct is much smaller, thus resulting in a much greater centrifugal force in the implanted circular construct.

As the velocity of the liquid metal in Figure A-2 is increased the scale of Figure A-2 will show less weight. Also, as the velocity of the liquid metal is further increased the pump and the tubing will become weightless and try to pull away from the scale. This action is further demonstrated with the construct in Figure A-3. However, there is a twisting force in the construct. It will be shown how these unwanted forces are eliminated.

When a mirror image of the device is attached, the twisting forces are cancelled. After we maximize the positive fore of one element we will attach a mirror image of the element to cancel all impeding forces as shown in Figure A-4.

The impeding forces consist of the frictional rotational forces of the moving liquid mass against the shell of the structure. Another type of force that is also corrected is the inertial forces cause by the clockwise and counter clockwise changes of mass direction.

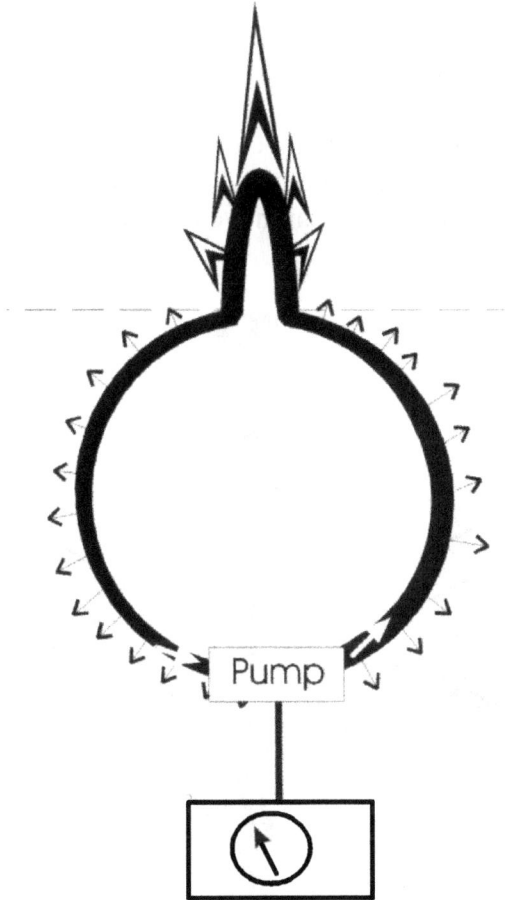

A-3

It can be seen in figure A-3 that by making the radius smaller and smaller the function MV^2/R acts more sharply and with greater amplitude.

Since the mass of flowing liquid metal is not a constant, but changes with flow, as follows:

(3) Mass of a Fluid = area * velocity * density, or a constant K * V
Using Kv for mass and substituting into MV^3/R gives:

(4) Force = KV^3/R
K is a constant, V is velocity, R = radius

Equation 4 demonstrates that velocity is much more important than mass or density of the liquid metal. This means almost any fluid can be used as long as the pump can move it through the structure quickly enough.

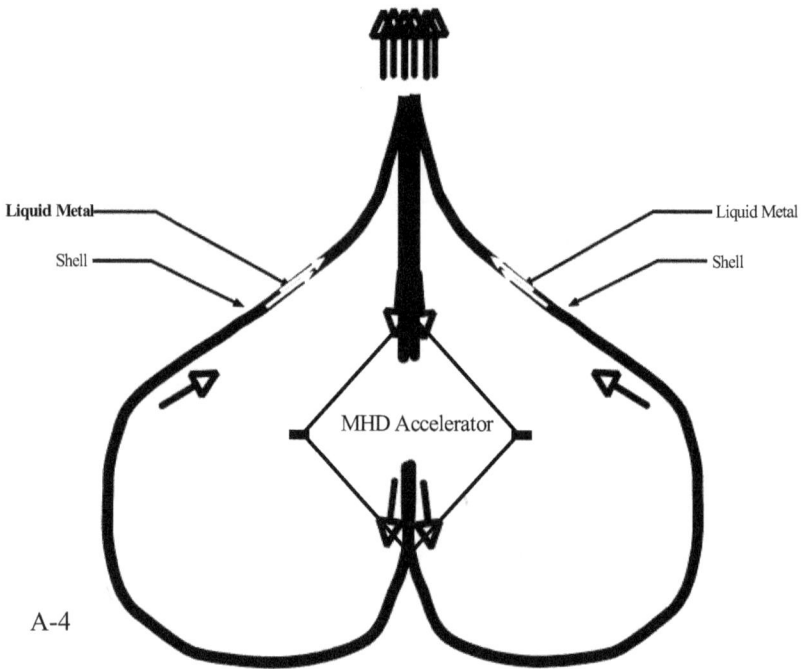

Liquid Metal
Shell
Liquid Metal
Shell
MHD Accelerator
A-4

Note that, it is unconceivable that the MHD Accelerator, structure and power source could left its own weight, but in space this system would be a viable means of propulsion. As you will see in Appendix B a more practical construct will evolve when internal momentum is used in conjunction with the unidirectional gyroscopic force of Appendix A.

APPENDIX B

Proof Closed System Propulsion Construct Using Liquid Metals Part 2

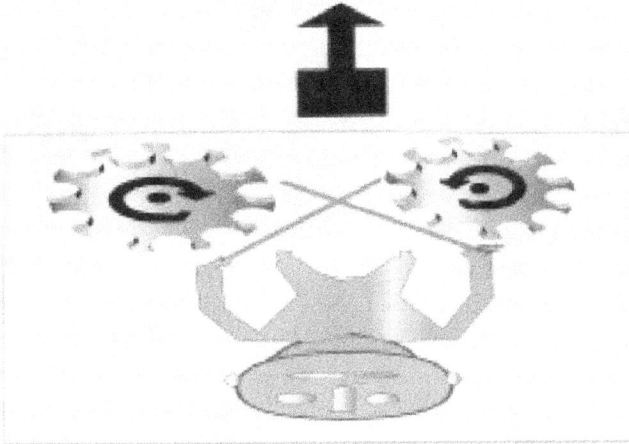

In this system a man is able to move in space and is propelled foreword by a rowing action. As the man moves and rows, the wheels spin in the direction shown. The meaning of this can be explained by a simple analogy. Imagine a man floating in space with two wheels of metal.

Let the man and the wheels remain stationary at this time. The man then reaches up and grabs the sides of the wheels and propels himself forward in space.

This results in the man moving forward as the two wheels spin and move in the opposite direction very slowly. His momentum is in the opposite direction.

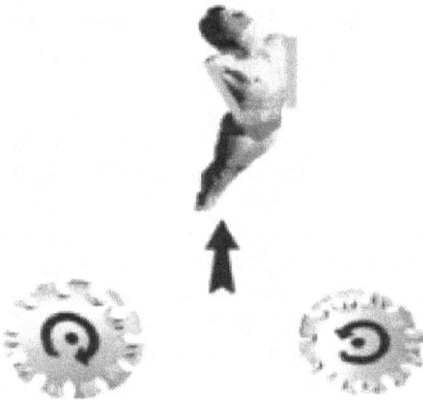

According to the law of Conservation of Momentum the momentum of the system is contained in three parts, the MV of the man, the MV of the wheels and the angular momentum (I W) of the two wheels. Also this law states that the sum of these parts is equal to zero.

This gives a simple equation as follows:

1. MVman + (MVwheels + IW) = 0

Assuming that some of the energy went into the angular momentum of the wheels instead of being propelled backwards gives the following:

2. MVman + IW > 0

The velocity of the system together is equal to the velocity of the man and is a vector greater than 0 in the forward direction as shown below.

I t follows then that the man and the wheels could be connected together and that the entire system would move forwards in space with a velocity as follows:

SYSTEM VELOCITY = [2*(Moment of Inertia of Wheel)*(Angular velocity of wheel)]/ [MASS OF MAN + 2*MASS OF WHEEL)]

As can be seem from the previous equation, the only way the system cannot move forward is if the angular velocity of the wheels is null.

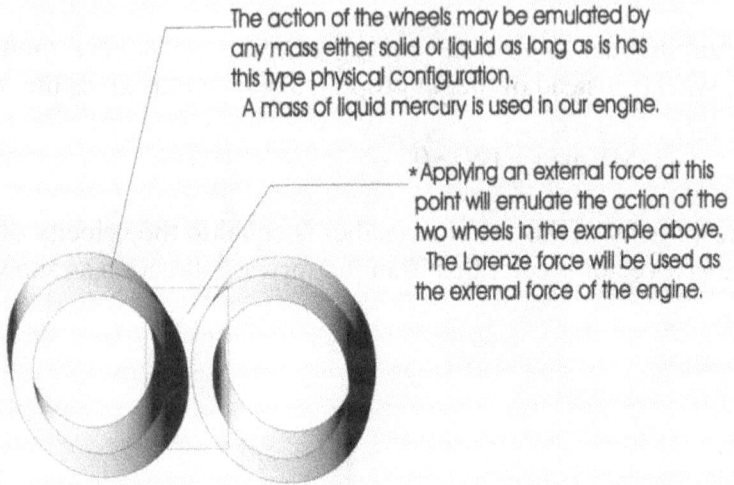

The action of the wheels may be emulated by any mass either solid or liquid as long as is has this type physical configuration.
A mass of liquid mercury is used in our engine.

*Applying an external force at this point will emulate the action of the two wheels in the example above. The Lorenze force will be used as the external force of the engine.

The Lorenze Force is used for the conversion of electrical energy to kinetic energy, maximum efficiency will be achieved with super conducting magnets or third generation magnets.

This system can be extended into a powerful and practical propulsion engine with the following analogy.

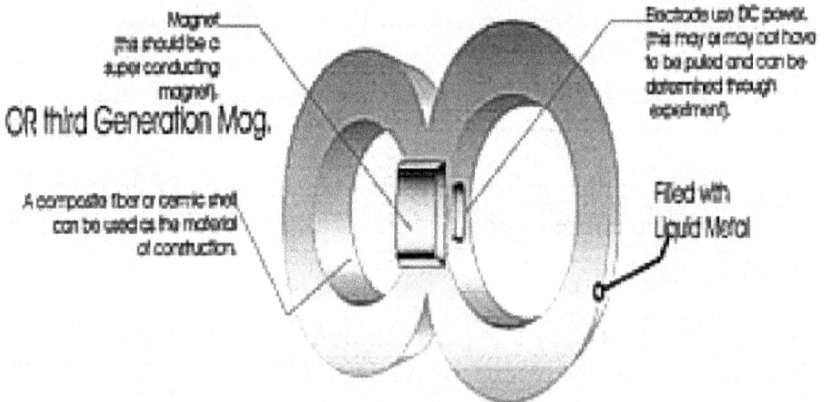

Magnet.
(this should be a super conducting magnet).

OR third Generation Mag.

A composite fiber or ceramic shell can be used as the material of construction.

Electrode use DC power. (this may or may not have to be pulsed and can be determined through experiment).

Filed with
Liquid Metal

Although the construct described above is useful and powerful when used in space it is only a partially complete system. A more powerful and practical propulsion engine is conceived by combining the functions described in appendix A.

The internal angular momentum method can be used in conjunction with the gyroscopic mode to improve the overall system power and efficiency. A completed system or construct referred to as a Closed System Propulsion Construct is illustrated in the following figure. Note that shape changes in the liquid metals direction that result in a propulsive force are accounted for in Appendix A while the conversion of angular momentum to linear momentum is described here.

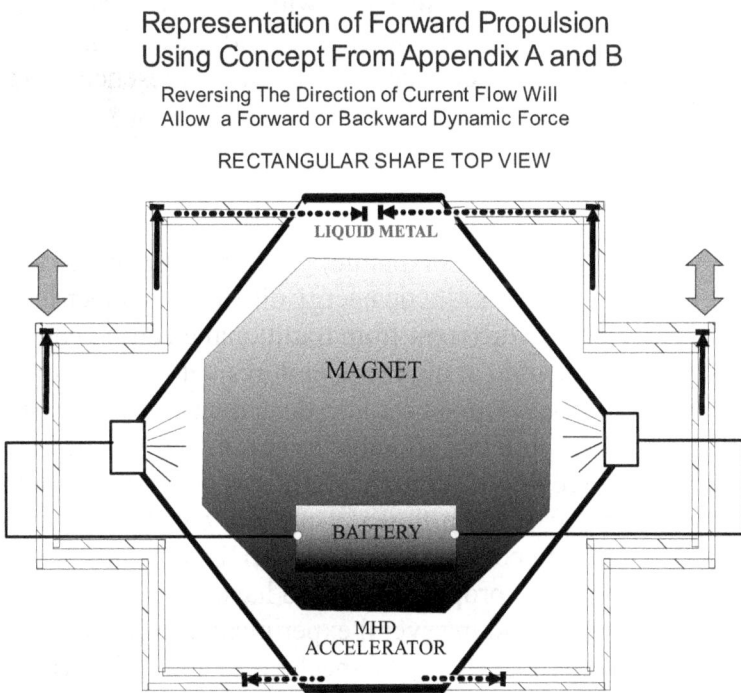

Representation of Forward Propulsion Using Concept From Appendix A and B

Reversing The Direction of Current Flow Will
Allow a Forward or Backward Dynamic Force

RECTANGULAR SHAPE TOP VIEW

LIQUID METAL

MAGNET

BATTERY

MHD
ACCELERATOR

Figure 5

All equations in Appendix A and Appendix B can be found in the text book **Serway PHYSICS:For Scientists and Engineers.**

Appendix C

Magnetohydrodynamics, Low Conductivity MHD Pumps, High Conductivity Liquid Metal MHD Accelerators/Pumps, Details of Basic Closed Loop Propulsion Construct Operation, Advanced Closed Loop Propulsion System

An MHD accelerator/pump is simply an MHD generator that is designed to operate is reverse. Where an MHD generator produces electricity when a conductive mass (solid, liquid, or gas) is forced through a magnetic field, the opposite is true with an MHD accelerator/pump. With an accelerator/pump a conductive fluid is accelerated when a current is passed through a conductive fluid in the presence of a magnetic field.

From Wikipedia:

"The MHD (magnetohydrodynamic) generator or dynamo transforms thermal energy or kinetic energy directly into electricity. MHD generators are different from traditional electric generators in that they can operate at high temperatures without moving parts. MHD was developed be-cause the exhaust of a plasma MHD generator is a flame, still able to heat the boilers of a steam power plant. So high-temperature MHD was developed as a topping cycle to increase the efficiency of electric generation, especially when burning coal or natural gas. MHD dynamos are the complement of MHD propulsors, which have been applied to pump liquid metals and in several experimental ship engines. The basic concept underlying the mechanical and fluid dynamos is the same. The fluid dynamo, however, uses the motion of fluid or plasma to generate the currents which generate the electrical energy. The mechanical dynamo, in contrast, uses the motion of mechanical devices to accomplish this."

An MHD accelerator/pump/propulsor is simply an MHD generator that is designed to operate in reverse. Where an MHD generator produces electricity as is the case when a conductive mass (solid, liquid, or gas) is forced through a magnetic field.

The opposite is true with an MHD accelerator/pump. With an accelerator/pump a conductive fluid is accelerated or pumped when a current is passed through a conductive fluid at right angles in the presence of a magnetic field. It will referred to as an MHD pump in the instance of low power and an MHD accelerator in instance of high power. This is because the specific applications differ.

Low Conductivity MHD Pumps

This type of device is generally use in low power applications where a steady state pumping action is desired. This includes slow flowing liquid metal cooling systems such as those used to cool electronic chips and nuclear reactors.

They may also be adopted in some high power systems, when a low conductive fluid such as sea water is being used. These are generally of two types: submarine propulsion and ion/plasma space propulsion. In any case the conductivity of the fluid and high power together demand that a restrictive geometry be used for the MHD accelerator.

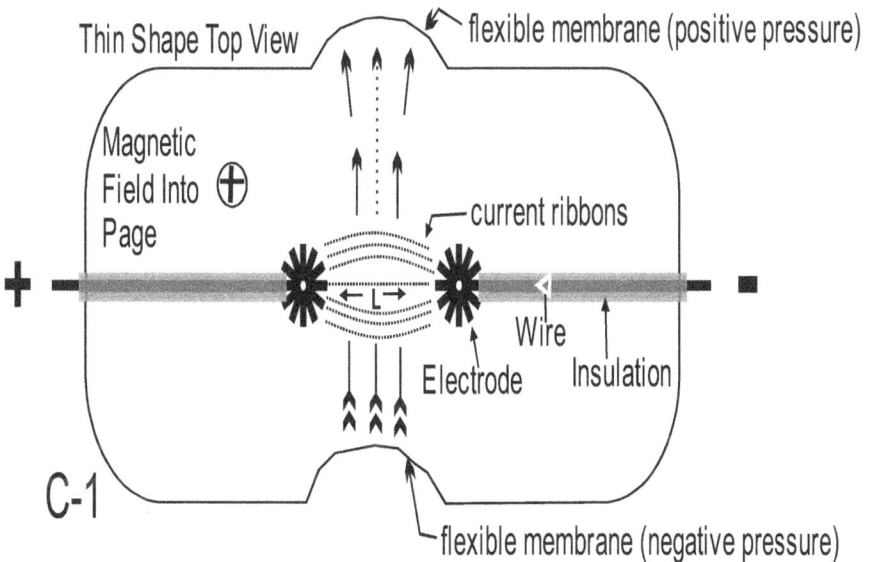

Thin Shape Top View — flexible membrane (positive pressure)

Magnetic Field Into Page ⊕

current ribbons

← L →

Wire

Electrode   Insulation

C-1

flexible membrane (negative pressure)

Typical MHD Pump Shown Above

High Conductivity MHD Accelerator

The low conductivity of the MHD pump in figure C-2 will operate normally for its applications. These particular applications, as been described earlier, are for cooling, steady flow at relatively low pressure and where the fluid is primarily for low velocity systems.

A fluid with high conductivity, high acceleration, and high power is required in our closed loop propulsion system. This is because the force produced in the closed loop propulsion system is directly proportional to the output of the MHD accelerator. As will be demonstrated below several special methods are required to produce the highest efficiency possible.

Using the normal geometry it can be seen in figure C-2 that most of the force is dispersed throughout the structure. By redefining the ordinary geometry of the structure in order to be used in a highly conductive fluid, we can constrain the current ribbons in a method that will pressurize the liquid metal.

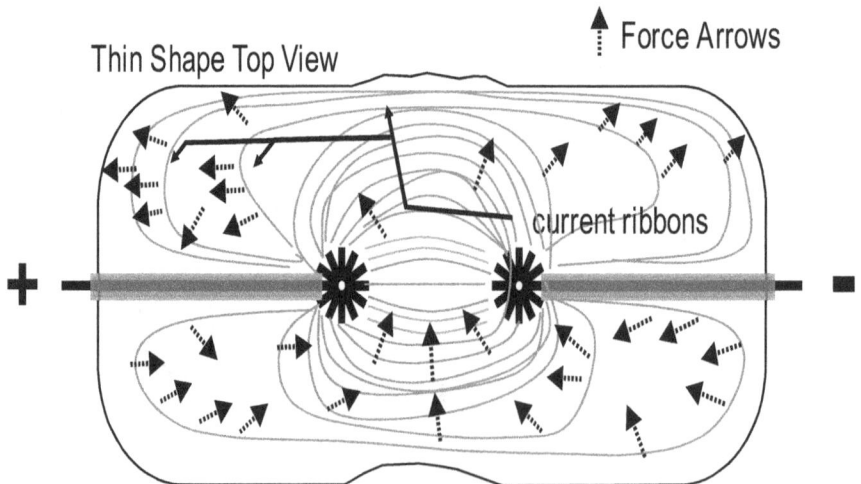

Thin Shape Top View

Force Arrows

current ribbons

+ − − ▪

C-2   as can be seen the focus of force is disbursed

High Pressure High Flow MHD Accelerator Structure

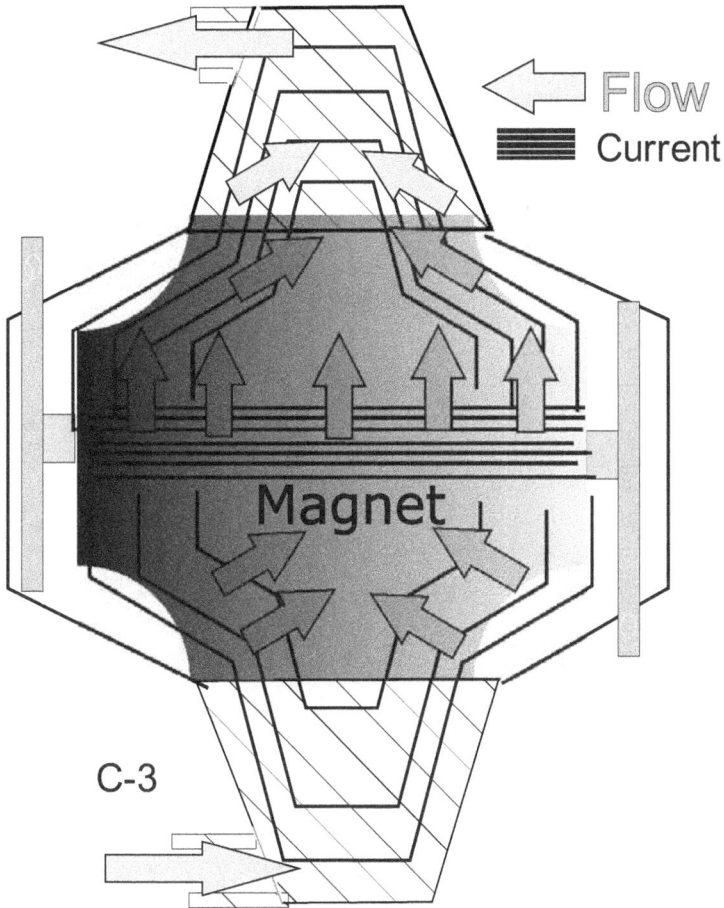

The illustration in Fig. C-3 is a practical high power high functioning MHD Accelerator. The structure should be as thin as practicable so as to allow a strong magnetic field to exist within the cavity. The cavity is filled with a liquid metal. The liquid metal is such that it should have both high conductivity and high mass density.

It is important to have high mass density since this is necessary to produce high force with changing mass direction. The MHD output of the pump is directly proportional to the current magnitude and the current magnitude is proportional to the conductivity of the liquid metal.

Details of Basic Closed Loop Propulsion Construct Operation

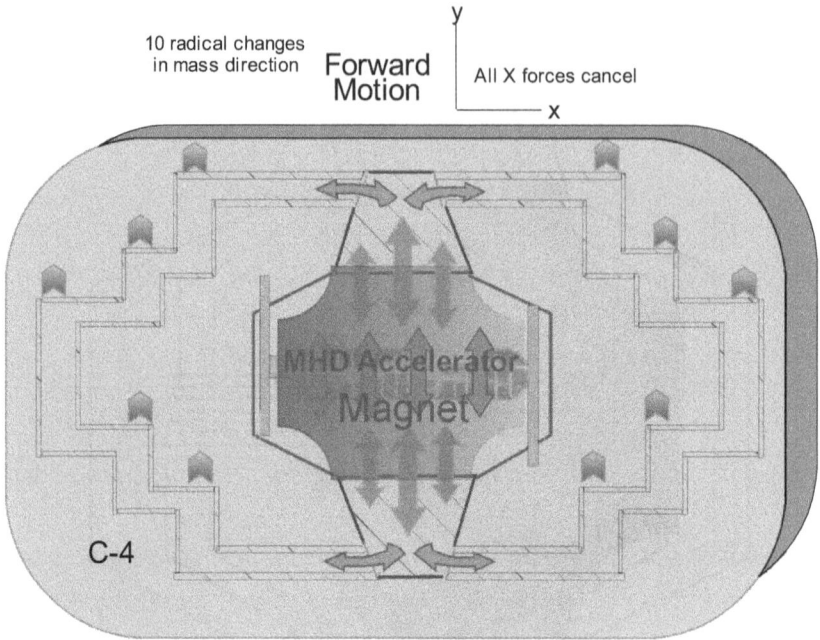

10 radical changes
in mass direction

Forward
Motion

All X forces cancel

MHD Accelerator
Magnet

C-4

From Wikipedia:

"Newtons Second Law can be stated as: The force acting on a body in a fixed direction is equal to rate of increase of momentum of the body in that direction. Force and momentum are vector quantities so the direction is important. A fluid is essentially a collection of particles and the net force, in a fixed direction, on a defined quantity of fluid equals the total rate of momentum of that fluid quantity in that direction.

Consider a mass m which has an initial velocity u and is brought to rest. Its loss of momentum is m.u and if it stopped in a time interval t then the rate of change of momentum is m.u /t.   The force F required to stop the moving mass is therefore F = m.u / t . Now if this is applied to a jet of fluid with a mass. Also in accordance with Newtons third law the resulting force of the fluid by a flowing fluid on its surroundings is (F).

The figure below illustrates this principle.

The fluid flowing in the pipe in the horizontal direction is forced to change direction at the bend such that its velocity in the original direction is zero. The reaction force on the pipe is F in the horizontal direction as shown."

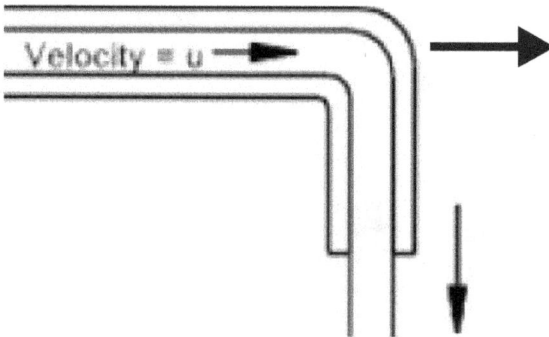

Velocity = u

As it can be seen from the above figure a force is exerted on the tube or duct by the flowing fluid. This force is directly proportional to the following: the velocity of the fluid, the density of the fluid, and the area of the tube or duct.

In figure C-4 the basic Closed Loop propulsion engine is shown. It is basically composed in two parts the MHD accelerator and the surrounding geometric structure. The geometric structure consists of two symmetrical sections the left side and right side. All forces on the left and right side of the structure balance and as a result no trust or dynamic force will result. Forces on the left and right of the structure involve dismemberment of the momentum and dissipation of the energy necessary to achieve the desired vectors for the desired vertical thrust or dynamic force.

Figure C-4 shows the operation of our Closed Loop propulsion engine in the positive Y direction. In this particular geometry of the Closed

Loop propulsion engine there are 10 radical changes in mass direction, mass directional change force is also covered in Appendix A.

All these forces are additive the other additive force is described in appendix B and is result of the angular momentum conversion.

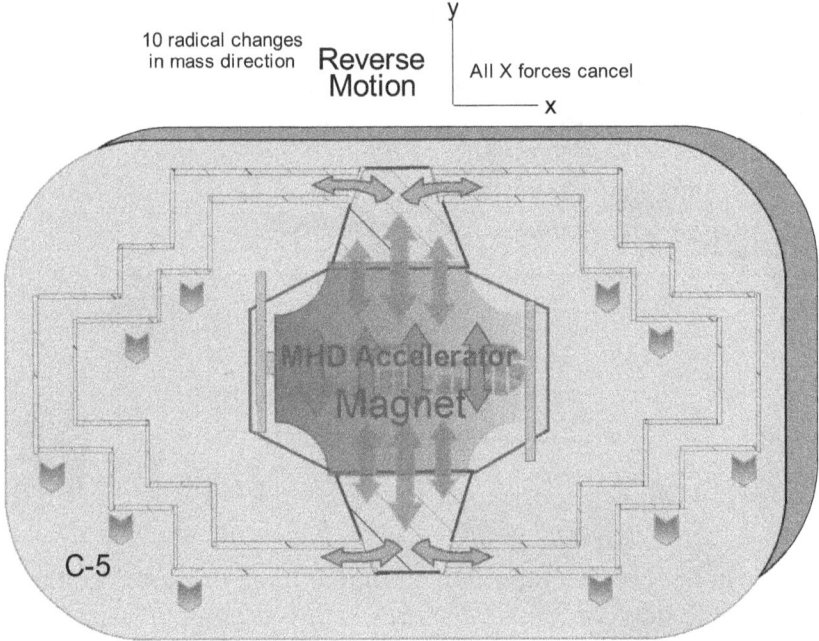

By reversing the current through the MHD accelerator, the direction of the closed loop propulsion force will also be reversed as shown in figure C-5.

Advanced Closed Loop Propulsion System

A more advanced system can be developed from the constructs in figures C-5 and C-6.This new construct will have very little increase in size and offer higher efficiency. It will also be capable of producing a multi-directional and multi-angular force. .

An advanced Closed Loop propulsion system is shown in Figure C-6. The ports or duct tracks, fluid port shapes that connect from the front to the back of the MHD accelerator, are doubled in duct or tube area allowing a two fold increase in mass per second from the original constructs. Doubling the area of mass flow increases the force or thrust by two.

This is a simple interpolation of the original construct. By dividing the original construct in half, adding another MHD accelerator and rotation studs for pivotal axis we have created a powerful and flexible Closed Loop Propulsion Engine. Is can be seen in the design of C-6 the engine has four wire connectors. By manipulating the currents magnitude and direction through these four connectors is possible to change the XY forces of the system. It is also possible to rotate around the pivotal axis to the left and right directions. This can be useful for satellites and corrections systems and also long-range probes where there is a need for small corrections at any given moment.

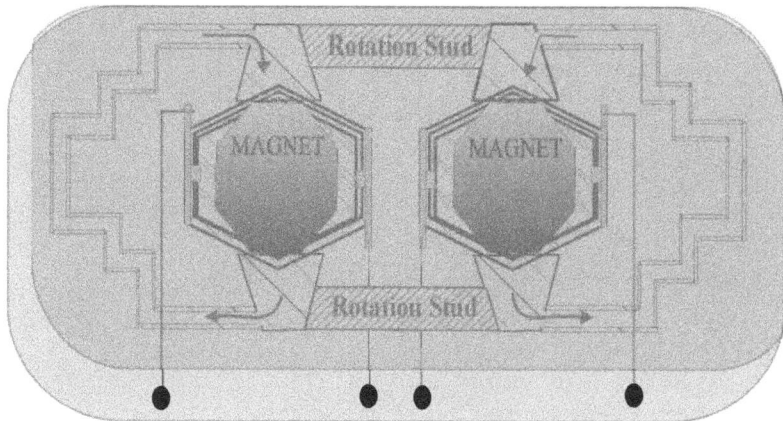

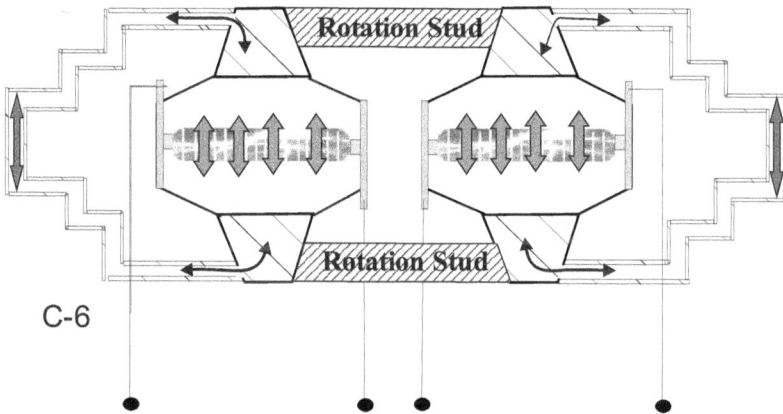

C-6

www.ingramcontent.com/pod-product-compliance
Lightning Source LLC
Chambersburg PA
CBHW060502210326
41520CB00015B/4059